Nívia Magalhães da Silva Freitas
Leonardo F. da Paixão
Nadia M. S. Freitas

Ideas on sustainability

Nívia Magalhães da Silva Freitas
Leonardo F. da Paixão
Nadia M. S. Freitas

Ideas on sustainability

Approaches to the complex approach

ScienciaScripts

Imprint

Cover image: www.ingimage.com

This book is a translation from the original published under ISBN 978-3-330-20020-3.

Publisher:
Sciencia Scripts
is a trademark of
Dodo Books Indian Ocean Ltd. and OmniScriptum S.R.L publishing group

120 High Road, East Finchley, London, N2 9ED, United Kingdom
Str. Armeneasca 28/1, office 1, Chisinau MD-2012, Republic of Moldova, Europe
Printed at: see last page
ISBN: 978-620-8-26699-8

"Indifference to the environment and connivance with our destruction".

Hans Alois

DEDICATION

To education professionals, particularly science teachers, who seek to innovate in their teaching.

(Nivia Magalhaes da Silva Freitas)

To the science teachers who strive to develop quality teaching.

(Leonardo Fernandes da Paixao)

To science teachers who value discussions on the environmental issues of our time.

(Nadia Magalhaes da Silva Freitas)

ACKNOWLEDGMENTS

To the students of the Magalhaes Barata State School of Primary and Secondary Education and the Anisio Teixeira School of Technological Education, located in the municipality of Belem, in the state of Para, who made it possible to carry out the research and write this book.

SUMMARY

The current capitalist economic system has been changing the dynamics, organization and structure of societies. In this context, various crises have arisen. The environmental crisis stands out in this context, with all sorts of interrelated events. Sustainability is a central issue in contemporary times and has therefore been the subject of studies and research in various fields of knowledge. The issue of sustainability has also had an impact on the educational field, through its consideration in teaching and learning processes. We therefore sought to answer the following research questions: in what terms do primary school students have a critical understanding of what constitutes sustainability? What dimensions are presented in the enunciation of ideas about sustainability? The research was based on a qualitative approach and we used a questionnaire containing one open-ended question as a data collection tool: What do you understand by sustainability? The analysis showed that the students' ideas about sustainability were more associated with concern for the environment - the environmental dimension of sustainability. We believe that this relationship - sustainability and environmental issues - is pertinent. Certainly, this apprehension was favored by the context surrounding the current environmental crisis. But other dimensions, even in association, were incorporated into the students' understanding of what sustainability is, entering the field of complexity. The issue of sustainability has emerged today as a topic to be prioritized in the classroom environment, particularly in order to present and discuss the implicit aspects of a development model that is based primarily on the reproduction of capital, depriving people of the conditions for a quality life and degrading the environment.

INDICE

1 INTRODUCTION

The transformations that have taken place in the social, economic and cultural spheres, based on the capitalist economic system, have been changing the dynamics, organization and structure of societies. In this context, various crises have taken shape. In this context, we highlight the environmental crisis, characterized by the following events: contamination and degradation of ecosystems, depletion of natural resources, climate change, loss of biological diversity, among others (MARQUES, 2016, PORTO-GONQALVES, 2012; SCHONS, 2012; VILCHES; GIL-PEREZ, 2013; 2015; 2016).

Until the middle of the 20th century, the planet seemed immense, and the effects of human activities on the environment were felt in certain places and, therefore, showed up
(NAVARRO; GARCIA, 2010). Today, however, it can be seen that these problems transcend borders, moving from the local to the global scale and thus occurring in various geographical spaces (WORLDWATCH INSTITUTE, 2012).

According to Leff (2002), in addition to an environmental crisis, we are experiencing a crisis of civilization, resulting, among other things, from the relationships established between human beings and nature, especially with regard to consumption, exploitation and alteration of natural environments and systems. People are even talking about the "[...] Anthropocene as a new geological period, marked by the fact that the major changes on the planet - and, very specifically, in the biosphere - are due to human action" (VILCHES; GIL PEREZ, 2011, p. 394).

Also, for Tristao (2004, p. 23), the "[...] environmental crisis is not just an ecological crisis but a crisis in the thinking of Western civilization". In this context, we can say the following:

> [...] most societies adhere to the competitive struggle for existence, believe in progress as economic and technological growth, overvalue individuality to the detriment of the possibilities of human interaction, solidarity, cooperation and so many other characteristics that make us human (VASCONCELOS et al., 2014, p. 247).

On the other hand, the environmental crisis is not restricted to a given locality, it is globalized, as Lobera (2008, p. 53) points out:

> Although the environmental impact of pre-industrial cultures was fundamentally restricted to the local environment, with the industrial revolution the consequences of pressure on the environment grew exponentially, surpassing the locality to reach global dimensions.

In the same vein, we have the observations of Quintana et al. (2011, p. 228), which are:

> For the first time in history, the degree of integration in which humanity finds itself is reaching levels of cause and effect never before experienced by human cultures. In this context, the environmental issue has emerged on a local and global scale, due to the growing environmental impacts generated by the dominant capitalist mode of production based on the unbridled use of natural resources, oblivious to the rhythms of nature's reproduction.

For his part, Loureiro (2000, p. 24) observes the following:

> The causes of environmental degradation and the crisis in the relationship between society and nature do not emerge solely from circumstantial factors or from humanity's perverse instinct, and the consequences of such degradation are not just the result of the improper use of natural resources; rather, they are the result of a set of interconnected variables derived from the categories: capitalism/ modernity/ industrialism/ urbanization/ technocracy.

According to Bigliard and Cruz (2007, p. 128), the following is evident:

> Today's society, characterized by the capitalist way of life and oriented towards consumption - has been treating natural resources as a source of raw materials for its consumption and perceiving the natural environment as a dumping ground for its waste.

According to Fonseca Neto, Freitas and Freitas (2017, p. 138), our society "[...] is engendered in the culture of consumption, alienated and alienating. Many are in conflict, and don't even know it, because we are confronted daily with two fundamental ways of being in the world: the way of having and the way of being."

Thus, we can consider that the environmental crisis, like all its developments and contributing elements "[...] is linked to the historical bases of the evolution of societies, notably when the capitalist model was adopted" (VASCONCELOS; CONCEIQAO;
FREITAS, 2012, p. 206).

Bauman (2005, p. 47) makes the following point:

> The 'problem of capitalism', the most glaring and potentially explosive dysfunction of the capitalist

> economy, is shifting from exploitation to exclusion [...] which today underlies the most obvious cases of social polarization, deepening inequality and increasing poverty, misery and humiliation.

In the same vein, Barbieri (2010, p. 148) observes: "[...] there are those who believe that economic growth is the source of the serious environmental and social problems observed in the contemporary world".

This state of affairs is characterized by unsustainable processes, which have governed the current neoliberal-capitalist model of productive development (BARQUERO; CREMONESE, 2006) . The fact is that, "[...] we are witnessing the crystallization of a marketized society, servile to the dictates of the development model, based on an anthropocentric view of the world" (FONSECA NETO; FREITAS; FREITAS, 2017, p.134).

In today's society, what prevails is the private appropriation of nature - nature as a commodity, associated with the ideology of individualism to the detriment of collectivism (BARENHO; MACHADO, 2007) . In this sense, Silva (2014, p. 46) says: "[...] what is hegemonic is a society based on individualism". But we have to consider that it is unacceptable to impose intolerable damage on a large contingent of the population under the pretext that they are the foreskin of progress (ECHEVERRIA, 2004; SACHS, 2001).

In addition, there are liabilities not only in the environmental field, but these extend to the social field, forming the following conditions: hunger, unemployment, poverty, social exclusion and violence (VILCHES; GIL-PEREZ, 2015). This state of affairs represents the depreciation of the human condition.

Thus, a paradigm shift is imperative "[...] from a technical and instrumental rationality to a social, open and cooperative rationality [...]" (FREITAS; MARQUES, 2016, p. 2). However, according to the same authors, such an attempt "[...] is not easy, given the generalization, the rooting of values and the social way of life contained in capitalist assumptions in contemporary society" (FREITAS; MARQUES, 2016, p. 2).

At this point, we have to consider Cavalcanti's considerations (2009, p. 160), namely: the "[...] use [...] of material and energy in doses 11
Excessive and growing depletion of environmental resources beyond their regenerative capacity obviously tends to make them less available for future generations". Therefore, it is necessary to establish "[...] the rational use of natural resources and the improvement of technical efficiency in terms of material transformations and energy use" (MARQUES, 2014).

This is because we will be considering the physical limits of nature, inscribed in the laws of thermodynamics, precisely in the Second Law (GEORGESCU-ROENGEN, 2012). And, on the other hand, we have to expunge the idea "[...] of common sense, for example, about the parsimonious use or economy of natural resources - even if important - [and] accept the idea of their finiteness on our planet" (BARBOSA; MARQUES, 2015, p. 1127-1128) and thus establish a prudent relationship in order to prolong human existence on the planet.

Sustainability is a central issue in contemporary times and has therefore been the subject of studies and research in various fields of knowledge (LEFF, 2016; LAYRARGUES; LIMA, 2014; MARQUES, 2016; VEIGA, 2010). Ultimately, the aim is to spread the assumption of sustainability, especially in development processes, so that we can guarantee a predictable future for society (LAURIE et al., 2016; NABAES; PEREIRA, 2016). The consequences of this perspective have also had an impact on the educational field, through its consideration in teaching and learning processes.

At this point, we must emphasize that education is a privileged way of spreading the assumption of sustainability. But not just any education, but one that is committed to forming citizens with a socio-environmental sensibility, in order to contribute to the cultural changes needed to shape a more balanced relationship between human beings and nature.

We therefore sought to answer the following research questions: in what terms do primary school students have a critical understanding of what sustainability is? What dimensions are presented in the enunciation of ideas about sustainability?

We believe that getting to know students' ideas about these issues is a key point for developing more enlightening, interesting and motivating teaching strategies, so as to enable students to learn in a socially meaningful way. On the other hand, we understand that improper notions of the term and its consequences make it impossible to adopt behaviors, attitudes, values, among other aspects, that will make our planet a socially and environmentally more suitable place for the aspirations of human societies.

This book is an expanded and updated version, in terms of theoretical references, of the research for the conclusion of the Biological Sciences degree course by the first author, supervised and

co-supervised by the second and third authors, respectively. This work is organized into the following sections, in addition to this Introduction: (i) "Discussions on environmental issues: some considerations"; (ii) "Sustainability and Sustainable Development: a necessary distinction"; (iii) "Methodological design"; (iv) "Ideas on sustainability: approaches to the complex approach"; (v)
"Final considerations".

2. DISCUSSIONS ON ENVIRONMENTAL ISSUES: SOME CONSIDERATIONS

Environmental issues have long been a concern for society. They have therefore featured on national and international political agendas (WATANABE-CARAMELLO; STRIEDER; GEHLEN, 2012). In the distant past, such issues were addressed, for example, in the following works by Darwin*: The Loves of plants and animals: romantic science and the pleasures of nature* (DARWIN, 1794) and *Temple of nature; or, origin of society* (DARWIN, 1804).

These publications portray fundamental aspects of environmental discussions. It is about recognizing nature for its intrinsic value, which can be safeguarded, and not as an object, a mere resource to be exploited and/or exhausted.

However, it was after the publication of Rachel Carson's book *Silent* Spring in 1962 that the environmental issue had a strong impact on society (CARSON, 2010). According to Lear (2010, p. 11**),** Carson's "[...] book initiated a transformation in the relationship between human beings and the natural world, and sparked the awakening of public environmental awareness".

Events and other publications were also responsible for highlighting the consequences of the intense use of natural resources. It is in this sense that the debates, the result of the major conferences and meetings of the United Nations, from the 1970s onwards, made public the prospective issues of the myth of unlimited economic growth, based on a base of finite natural resources that are vulnerable to capitalist expansion (LIMA, 2009).

We highlight the United Nations Conference on the Human Environment, held in Stockholm in 1972, the first major meeting of heads of state organized by the United Nations (UN) to address issues related to environmental degradation. This event, in its debates, led to the environment being considered a priority issue on the international agenda (UNEP, 2012). Also at this conference, according to Leff (2002, p.16), "[...] the limits of economic rationality and the challenges of environmental degradation to the civilizing project of modernity were highlighted".

In turn, the publication of the report *The Limits to Growth* (MEADOWS et al., 1973) also contributed to highlighting the need to redefine the intensity of the use of natural resources. The perspective was to present the idea of limits, in the sense that nature imposes limits on society, notably on unlimited economic growth on the basis of limited resources.

Another important milestone in the environmental field was the publication of the *Our Common* Future report. This report was drawn up by the UN World Commission on Environment and Development under the chairmanship of Gro Harlem Brundtland, hence its name. In this report, the concept of Sustainable Development was presented and incorporated into the political agenda, as one capable of "[...] meeting the needs of present generations while taking into account the needs of future generations" (COMMISSION, 1991, p. 46).

Also noteworthy is the United Nations Conference on the Environment (UNCED), known as the Rio 92 Conference, held in Rio de Janeiro. According to Afonso (2006, p. 20), UNCED is "[...] considered a historic milestone in the discussion of environmental issues because it was the first meeting to discuss the political, social and economic aspects of environmental problems".

During this event, heads of state and government approved Global Agenda 21, the content of which asserts the need for changes in development patterns. Global Agenda 21 also states that education is one of the main instruments for promoting sustainable development, in the following terms:

> [...] education is of fundamental importance in promoting sustainable development and increasing people's capacity to address environmental and development issues [...]. Education is also fundamental to providing environmental and ethical awareness, values and attitudes, techniques and behaviors that are in line with sustainable development and that foster effective public participation in decision-making (CONFERENCE, 1997, p. 533-534).

Thus, with the aim of contributing to discussions on sustainability, and collaborating in the fulfillment of Global Agenda 21, the United Nations Educational, Scientific and Cultural Organization (UNESCO) launched the United Nations Decade of Education for Sustainable Development (DESD), which was instituted in December 2002 by the UN General Assembly, covering the period 2005-2014 (UNESCO, 2005). This educational proposal, in our view, assumes sustainable development as a vision of the future, on which humanity needs to base its objectives, according to Veiga (2005).

According to UNESCO (2005), the aim of the Decade was to highlight the importance of integrated actions to guarantee sustainable development standards. Furthermore, Education for Sustainable

Development (ESD) plays an important role, particularly in the correct perception of environmental and social problems.

On the other hand, we can consider the following:

> Education for Sustainable Development plays an important role because it is geared towards fostering a correct perception of socio-environmental problems, as well as encouraging the formation of a critical, autonomous, creative citizen who is aware of their (fragile) human condition, capable of understanding the complexity of the natural and social world and thus presumably prepared to face uncertainties and make decisions (CARDOSO et al., 2010, p. 8).

According to the same authors, ESD "[...] seeks to foster appropriate attitudes and behaviors, both individually and collectively, for the construction of sustainable development".
(CARDOSO et al., 2010, p. 8)

It is therefore important that, at the end of the DEDS period, we can (re)understand its real repercussions. Furthermore, we believe that this proposal is timely, given the scenario shaped by the environmental crisis; and, in this sense, it is necessary to make efforts to extend the educational approaches that have sustainability as their centrality.

Thus, in the context of the research work presented in this book, we believe that one of the ways to consider the repercussions of the environmental crisis is precisely to understand the ideas that students have about the term "Sustainability". Including considering its multidimensionality.

3 SUSTAINABILITY DEVELOPMENT SUSTAINABLE: A NECESSARY DISTINCTION

The terms sustainability and sustainable development have been used interchangeably. However, there is a body of work that indicates that these terms have different origins, definitions and understandings. So, in this chapter, we propose to present, albeit briefly, some of these distinctions.

Therefore, we can say that sustainability seems to have two origins, namely: (1) biological (branch of ecology), which refers to the ability of ecosystems to recover and reproduce (resilience) as a result of anthropogenic or even natural aggressions; (2) economic, adjectivizing the term development (sustainable development) (NASCIMENTO, 2012; VEIGA, 2010).

The term sustainability has been added to the most diverse fields of knowledge. And because it has different meanings and values, the term sustainability has become a polysemic concept (FAUSTINO; AMADOR, 2016). However, this does not exempt it from being a term that is difficult to define, since it is one of the concepts whose meanings are insufficient (VEIGA, 2010).

Given the profuse use of the term sustainability "[...] it seems to have become a kind of current mantra" (CAVALCANTI, 2012, p. 35). Even the notions of sustainability have served [...] Greeks and Trojans when they want to express vague ambitions of continuity or perpetuity. All [notions] refer to the future" (VEIGA, 2010, p.12). Ultimately, it can appear to be mere nominalism (VACARRO, 2013).

In any case, it is in Boff's conceptualization that we find significant elements that clarify what sustainability is:

> Sustainability is all about maintaining the energetic, physical and chemical conditions that sustain all beings, especially the living Earth, the community of life and human life, with a view to their continuity and meeting the needs of present and future generations, in such a way that natural capital is maintained and enriched in its capacity for regeneration, reproduction and coevolution (BOFF, 2012, p. 107).

In these terms, we can characterize it as a "[...] systematic, complex concept. And, as such, it comprises distinct dimensions which, however, have interdependent relationships or, in other words, intersect and fertilize each other" (FONSECA NETO; FREITAS; FREITAS, 2017, p.135). The same authors add that sustainability must be understood as a fundamental quality of development.

At this point, we should point out that the "[...] idea of sustainability is taking shape and gaining political expression in the adjective development, as a result of the perception of a global environmental crisis" (NASCIMENTO, 2012, p. 52). So, in the same vein, we can say that the concept of Sustainable Development arises from studies, reflections and discussions on the relationship established between human beings and nature, which, over time, has proved to be destructive.

The notion and discourse on Sustainable Development has emerged to take the place of the discourse on economic development. This is because economic development has proved to be inefficient, since it has not been able to fulfill the promises of progress and social welfare (LIMA, 2009). On the contrary, social inequalities have deepened (BARQUERO; CREMONESE, 2006; CACCIA BAVA, 2007; KLIKSBERG, 2002) and environmental degradation has intensified

(RATTNER, 2012; ZIBETTI, 2006).

Changing the unsustainable conditions of current development processes to a Sustainable Development perspective will require considering development not only through the lens of Gross Domestic Product (GDP) growth, but mainly from the perspective of generating social opportunities (BUARQUE, 1999; SEN, 2010).

According to Sachs (2005), development models that focus on economic aspects and the naturalization of the problems arising from them have proved to be inefficient. This is because it is clear that these problems are not only not being solved, but are also intensifying. This has contributed to the triggering of various crises, such as social and environmental, among others.

Economic growth is necessary, but it is not enough in *itself* to promote development (CARNEIRO; BAGONI, 2012). Therefore, it is "[...] important that the economic factor can interact with socio-cultural and environmental aspects [...]" (FREITAS, 2008), becoming an instrument for improving all aspects of life, from a present and future perspective.

The multidimensional understanding of the assumption of sustainability as part of the concept of development needs to be grasped by society so that a paradigm shift can be undertaken: from economic development to sustainable development, incorporating social, environmental, cultural and other dimensions, in other words, emphasizing the different dimensions of sustainability in development.

In fact, we have seen advances in this understanding, as highlighted by Zoottis et al. (2008, p 35):

> The concept of sustainable development [...] has come to encompass four basic dimensions: the ethical, which involves solidarity with future generations; the temporal, through precaution and long-term planning; the social, through the reduction of inequalities; and the practical, which recognizes the need to change consumption habits and behaviour.

In the same vein, Jacobi (2003, p. 194-195) ponders: "[...] the notion of sustainable development leads to the necessary redefinition of human society/nature relations and, therefore, to a substantial change in the civilization process itself".

Corroborating this idea, we can mention the following:

> The need therefore arises to (re)construct the concept of development from a perspective that considers the economic, social, cultural, political and environmental aspects, with the aim of bringing about transformations in society's social and mental structures (FREITAS, 2008, p. 21).

Thus, from all that has been listed above in this chapter, we can understand, in agreement with Roloff (2011, p. 38), that Sustainable Development represents "[...] an attempt to overcome the limitations that technology and social organization impose on the environment, including economic and social variables".

The fact is that it makes sense to discuss the current development model, since it has caused society countless liabilities. And overcoming such a state of affairs will require betting on differentiated and combined approaches that promote development supports oriented towards the foreseeable future for humanity.

4 METHODOLOGICAL DESIGN

The research was based on a qualitative approach (OLIVEIRA, 2014). Godoy (1995, p. 21) states the following:

> [...] qualitative research occupies a recognized place among the various possibilities for studying the phenomena that involve human beings and their intricate social relationships established in various environments.

Furthermore, this type of research favors the understanding of reality and is more appropriate when you want to work with the universe of meanings, values and beliefs (MINAYO, 2008).

The research was carried out on the premises of the Magalhaes Barata State Elementary and Middle School and the Anisio Teixeira Technological Education School, both located in the municipality of Belem, in the state of Para. It involved 17 students who agreed to take part in the research (accessibility sample). The data collection instrument used was a questionnaire containing one open question: What do you understand by sustainability?

The research subjects, identified by the initial letters of their name/surname, were invited to answer the question in writing, after signing the Free and Informed Consent Form (FICF), a condition for complying with the ethical aspects of the research (BRASIL, 1996). The students' answers (data) were analyzed using the principles of content analysis proposed by Bardin (2011).

5 IDEAS ON SUSTAINABILITY: APPROACHES TO THE COMPLEX APPROACH

Life as it is today is part of the field of complexity - multiple domains make it up. So, in order to deal with the most critical issues that we face today, such as the environmental crisis, which has manifested itself with a whole range of interrelated problems, namely: loss of biodiversity, loss of biodiversity, loss of biodiversity, loss of biodiversity, loss of biodiversity, loss of biodiversity.
biodiversity, deforestation, water and soil contamination, erosion, desertification, climate change, ozone depletion, depletion of natural resources, among others (LEFF, 2009), it is necessary to consider the assumption of sustainability in all its multidimensionality.

It is in this context that this study was able to highlight, considering the elements pointed out by the students, three categories, which in reality refer to the dimensions of sustainability, namely: (a) environmental sustainability; (b) environmental and economic sustainability and (c) environmental and social sustainability. Note the combination of the dimensions of sustainability.

In presenting their ideas about sustainability, the students incorporated different dimensions, creating more complex understandings. The results and discussions presented here have been organized into three subsections, starting with the presentation of the students' ideas and then discussing them in the light of theoretical references from the field of sustainability.

a. Environmental sustainability

Environmental issues are currently at the center of concerns about sustainability. This is because the pressure on natural resources, imposed by the economic-productivist development model, is growing, especially in order to increase consumption, the basis of this model (Zacarias 2009). This scenario has led to the overexploitation of natural resources, jeopardizing the sustainability of the planet.

So, in this category, we grouped the answers referring to sustainability, with an emphasis on the environmental dimension, into two thematic blocks, in order to facilitate discussions. The first thematic block concerns recycling and the second refers to the issue of environmental preservation.

Block 1- Recycling

1. "Sustainability helps to benefit nature by taking objects and recycling them to improve the environment" (SC);
2. "And the method used for recycling in the environment" (AJ);
3. Sustainability [...] seeks new forms of interaction between man and the environment, with the 3Rs as its central idea: reduce, recycle and reuse (BC);
4. "[...] how to clarify recycling ideas [...]" (LP).

With regard to the three excerpts highlighted above, we can say that there is a lot of talk about recycling. However, we must bear in mind that intervention in unsustainable processes in today's society does not lie exclusively in the use of recycling as a way of ensuring an adequate relationship with the environment or even resolving the situation. Waste production is currently on the increase and stems from consumption, which starts a cycle that links the exploitation of natural resources to various economic and social aspects.

In reality, there are many factors that can be blamed for the emergence of environmental problems, which have created unsustainable conditions. Among these factors is consumption. Thus, we can mention the following:

> Consumption is the primary interaction between living beings and the environment, whether through the use of soil as a substrate and source of nutrients, or through the use of fossil fuel reserves as a source of energy (MELLO, 2009, p. 111).

However, in the current context of degenerative capitalism (MELO, 2006), it is clear that, as previously pointed out, "[...] the organizing factor of contemporary society lies in the sphere of consumption and not production" (ZACARIAS, 2009, p. 119), which is a triggering aspect of the growing despoliation of natural resources.

The majority of students attached importance to recycling in the context of sustainability. This points to the need for teaching approaches that promote critical reflection on "[...] the cultural values of consumer society, consumerism, industrialism, the capitalist mode of production [...]" (LAYRARGUES, 2002, p. 179).

This is because the search for environmental sustainability is related to the rethinking of the flows of energy and material that pass through the economic system (VEIGA, 2010). Today, as a result of the current environmental crisis, we are certain that we need to establish another relationship with the environment, an ecologically prudent one (SACHS, 2001), in order to guarantee a predictable future for humanity.

Block 2 - Environmental preservation

1. "It's a way of being aware of [...] nature, not damaging it" (IA);
2. "And the ability of human beings to interact with the world while preserving the environment" (CS);
3. "Sustainability is understood as a process of raising society's awareness of the need to preserve the environment [...]" (LT);
4. "It's all about preserving the environment" (LS);
5. "Sustainability is the word of the day: we must preserve nature. Sustainability for me is preservation" (ES);
6. "It's a way of life in which elements of nature are used without harming them" (FG);
7. "It is any action [...] aimed at preserving the environment [...] and using the resources that nature makes available for our sustenance in a conscious way" (DB).

The excerpts above, in some way, reflect the concern about environmental issues in society, in terms of the depletion of environmental resources. In fact, the current development model has caused serious environmental problems, creating the current environmental crisis as a result of the overexploitation of natural resources, among other aspects. This certainly reveals the unsustainability of this model, also in its environmental aspect (VASCONCELOS; CONCEIQAO; FREITAS, 2012).

Supporting this idea, Bigliard and Cruz (2007, p. 128) state that today's society, as a result of the adoption of the economic model, has a productionist "[...] has been treating natural resources as a source of raw materials for its consumption and understanding the natural environment as a deposit for its waste". So, the "[...] first step to reversing the planetary environmental crisis is to change our patterns of production and consumption [...]" (SERRAO; ALMEIDA; CARESTIATO, 2012, p. 104).

At this point, we must highlight the difference between the terms conservation and environmental preservation:

> Both are aimed at the sustainability and continuity of natural resources. However, preservation imposes that the area or ecosystem cannot be altered, while for conservation it is possible to reconcile educational activities or support within a Conservation Unit, a reserve, a park, etc. (ALVES et al, 2011, p. 5);

In the same vein, we have the following understanding:

> Conservation is the management of human use of organisms and ecosystems in order to guarantee the sustainability of this use. In addition to sustainable use, conservation includes the protection, maintenance, rehabilitation, restoration and

> enhancement of (natural) ecosystem populations (DIEGUES, 2000, p. 1).

However, regardless of whether or not the terms are used correctly, we can consider it plausible for students to think about the need for rational use of natural resources. In any case, it must be recognized that discussions about environmental sustainability are central when thinking about a presumed future for humanity. This is because the "[...] irreparable environmental impact of economic growth on nature's resources and sinks" (CAVALCANTI, 2012, p. 35).

b. Environmental and economic sustainability

The students here are referring to two dimensions of sustainability. Earlier, we mentioned environmental sustainability. Let's now look at the understanding of economic sustainability, in the terms of Fonseca Neto, Freitas and Freitas (2017, p.136): "We understand economic sustainability, basically, as the organization of material life".

We understand that the economic component of the development process is a fundamental condition, because according to Sachs (2001), it would be very difficult to redistribute goods and income in a stagnant economy. However, if it were possible, it would be morally repugnant to base all social value on a market price (SANTOS, 2006).

Thus, for Fonseca Neto, Freitas and Freitas (2017, p.136), "[...] one of the limits imposed on a market economy lies in considering the human component, the social quality of development". It's true that we can't do without the economic component, but its focus should be on meeting people's needs in terms of quality of life. Quality of life is multidimensional, as it encompasses work, health, housing, leisure, education, etc.

So, having done that, let's take a look at the students' answers, which combine the environmental and economic dimensions in a single block. The emphasis is still on the environmental dimension, as can be seen in the following excerpts:

> 1. "Human growth without harming nature, preserving it" (CS);
> 2. "They are means of production that, when produced, do not harm the environment, involve the economy, education, etc." (IA);
> 3. "It would be a way of developing without harming nature, for example, factories that use solar energy don't pollute as much as factories that obtain energy from burning fossil fuels" (DF);
> 4. "Sustainability is a way of reusing materials [...] this idea contributes a lot to the environment, which will be less exploited, and also reduces costs" (FP).

The answers given above reveal a broader approach, especially with regard to the concern to combine economic growth and environmental protection, although this combination seems remote (VECCHIATTI, 2004).

The economic aspect is highlighted by the students in the following terms: growth, production, development, cost reduction. Thinking about development processes (growth) with environmental concerns stems from the realization that there are limits to both, notably due to the understanding that the production process is based on a finite base of natural resources (CECHIN, 2010).

There is a need to reconcile, if possible, development/growth with the (rational) use of natural resources. This is because we are experiencing a mode of production (capitalist) that has led to the depletion of resources and the inability of the environment to be resilient (RICKLEFS, 2010; VEIGA, 2005).

It is important that, in the classroom environment, we are critical, problematizing the reality of a development process that insists on the degenerative appropriation of natural resources. And so, to lead us as follows:

> [...] no longer be carried away by a romantic and naïve vision of the environment or by the imaginary achievement of environmental sustainability based on the idea of a balance between economic reasons and the environment (BARBOSA; MARQUES, 2015, p. 1131).

We have to consider that being guided solely by economic interests is no longer acceptable in this century, given the current environmental crisis. Thus, a new orientation is needed for the different fields of society, the adoption of the assumption of sustainability (PEDUZZI; VILCHES;
GIL-PEREZ, 2014)

The aim is to draw attention to the need to avoid the process of extinction of the human species, to which we are subject, precisely because of the exhaustion of a large part of the ecosystems that make up the biosphere (VEIGA, 2010). These approaches need to be disseminated in the school environment, presenting the various contexts that imply unsustainability, with a view to forming critical subjects who challenge the current scenario.

1.3 Environmental and social sustainability

Here, the students brought up another dimension of sustainability - the social dimension, in association with the environmental dimension, already discussed above. Thus, we can consider social sustainability, as Fonseca Neto, Freitas and Freitas (2017, p.135), based on (SEN, 2010), put it:

> [...] as the permanence and long-term continuity of the efforts made to achieve development based on broadening the choices that people must have in order to lead a life according to their needs (material and/or non-material) and interests.

In this sense, we corroborate Veiga's statement (2010, p. 37), "[...] it would be simply ridiculous to think of sustainability in the face of the supreme humiliation of the human species itself, the deprivation of its freedoms and the stupidity of its killings". In reality, "[...] society currently needs to perceive reality from a humanizing perspective, sensitive to social and environmental issues [...]" (VASCONCELOS et al., 2014, p. 246). This means that development objectives must be aligned with society's social demands in order to constitute a genuine development process with social quality.

Considering the introductory aspects of this subsection, we can consider the last analytical category. This is the association of two dimensions of sustainability: environmental and social, presented in a single block.

In the students' understanding, both contribute to the establishment of sustainability in its most general aspect:

> 1. "It's a process for the world to have the advantages of being able to reduce the irregularities in our environment [...] a new way of living, without pollution and without heating up our climate. So that we can have a good quality of life" (C.T);
>
> 2. "An act of awareness is the preservation of the environment in which we live, bringing [...] education to the environment. Sustainability makes us understand that we must preserve, take care of [...] in order to continue our own life, a healthier life (SL).

In fact, improving the quality of life is one of the global imperatives, recognizing that the planet is being subjected to unbearable pressures that are detrimental to its own development, the environment and people's living conditions (SUNKEL, 2001).

Thinking about development means worrying about the human dimensions of sustainability, in other words, considering that the ultimate aim of development is to achieve a real improvement in people's quality of life (ZAPATA, 2005). It is not the market, not the economy, not production that matters or should matter, but people.

How can we think about sustainability when we face unemployment, housing shortages, slums, precarious health and environmental sanitation infrastructure, among other aspects (BARQUERO; CREMONESE, 2006)? According to Boff (2012, 19-20), the

"[...] the sustainability of a society is measured by its capacity to include everyone and guarantee them the means of a sufficient and decent life". In these terms, development must be at the service of social objectives (SACHS, 2005). Thus, the social dimension is incorporated into the broader context of sustainability.

We must emphasize, in agreement with Fonseca Neto, Freitas and Freitas (2017, p.138), that the current social and environmental situation in society "[...] is compromising the prolongation of life for all living beings on our planet". And, according to the same authors, it is necessary to understand the aspects involved in the current stage of development, in the scenario that is configured as an environmental crisis, precisely in the sense of literacy for sustainability.

Education is a privileged way of launching ourselves into an in-depth and judicious reflection on social and environmental problems and of setting out formative paths for considering the assumption of sustainability and, consequently, for the constitution of critical citizens in the face of the current social and environmental scenario, especially in the understanding that the current relationship between human beings and nature has proved to be damaging for both man and nature.

6 FINAL CONSIDERATIONS

Throughout the text, we draw attention to the discussions related to the theme of sustainability, especially its many dimensions and its place within the framework of situations that are characterized as unsustainable. And, consequently, we address the aspects that make up the current environmental crisis, that is, the harmful relationships established between human beings and nature, especially with regard to consumption, exploitation and alteration of natural systems, and also the consequences for the social field.

These situations have put the existence and prolongation of life on the planet, as we know it today, at risk. We point to capitalism as a degenerative aspect of the development process, mainly because it is based almost exclusively on an economic-productivist perspective.

We point out that sustainability represents a complex network of dimensions. In this context, we would like to highlight the dimensions that make up the sustainability trifecta, namely: social, economic and environmental, to limit ourselves to the dimensions most frequently mentioned in the literature.

The analysis showed that the students' ideas about sustainability were more associated with concern for the environment - the environmental dimension of sustainability. We believe that this relationship - sustainability and environmental issues - is pertinent. Certainly, this apprehension was favored by the context surrounding the current environmental crisis. However, other dimensions, even in association, have been incorporated into the students' understanding of what sustainability is, entering the field of complexity. It's true that we live in a world permeated by complexity. Multiple dimensions make up the different objects that are in focus in today's society. This is no different for teaching objects.

And understanding students' ideas on a contemporary topic such as sustainability can help us to avoid teaching and learning approaches that do little or nothing to critically unveil a given situation and/or object. We need to go beyond the mere reductionist transposition of content and incorporate more complex analysis, especially in the relationship between human beings and nature.

The issue of sustainability has emerged today as a topic that should be highlighted in the classroom environment, particularly in order to present and discuss the implications of a development model that is based primarily on the reproduction of capital, depriving people of the conditions for a quality life and degrading the environment.

Dealing with the subject of sustainability, in all its many aspects, does not mean inserting oneself into a fashionable discourse, an aseptic trend. On the other hand, it means considering sustainability as an essential quality for building societies that seek, on this basis, more benign development processes.

To help shape a possible future and
It is essential for everyone to be able to understand the destructive aspects involved in the current development process. And also to incorporate the multiple dimensions that make up sustainability into their analysis. This is because, in In contemporary times, sustainability has become a new value (FREITAS, 2012; VEIGA, 2010), especially when we consider the fearful scenario that poses the possibility of the

human species heading towards its own annihilation, in the face of the exhaustion of natural resources that is peremptorily announced.

REFERENCES

AFONSO, C. M. **Sustentabilidade**: caminho ou utopia. Sao Paulo: Annablume, 2006.

ALVES, F. G. The return to the metanarrative as a defense against the ethnocentrism of regional ontologies. In: Congresso Internacional de Historia: Cultura, sociedade e poder, 4. Universidade Federal de Goias, Jatai, September 23-25, 2014. **Electronic proceedings**... Available at: <http://www.congressohistoriajatai.org/2014/anais2014.html>.
Accessed on Jan. 20, 2017.

BARBIERI, C. J. et al. Innovation and sustainability: new models and proposals. **RAE,** Sao Paulo, v. 50, n. 2, p. 146-154, 2010.

BARBOSA, L. C. A.; MARQUES, C. A. Environmental sustainability and thermodynamic postulates in the light of the work of Nicholas Georgescu- Roegen. **Revista Eletronica em Gestao, Educagao e Tecnologia Ambiental**, Santa Maria, v. 19, n. 2, p. 1124-1132, 2015.

BARDIN, L. **Content analysis**. Lisbon: Edigoes 70, 2011

BARENHO, C. P.; MACHADO, C. R. S. Contributions of Marxism and ethnoecology to the study of socio-environmental relations. In: International Marx-Engels Colloquium, 5. Campinas, Sao Paulo, 2007. **Electronic proceedings ...** Available at: <http://www.unicamp.br/cemarx/anais_v_coloquio_arquivos/arquivos/comunicacoes/gt2/sessao3/Cintia_Barenho.pdf>. Accessed on: Nov. 16, 2015.

BARQUERO, M.; CREMONESE, D. Introduction. In: BARQUERO, M.; CREMONESE, (Org.). **Social capital:** theory and practice. Ijui: Inijui, 2006. p. 7-17.

BAUMAN, Z. **Identidades**: entrevista a Benedetto Vecchi. Rio de Janeiro: Jorge Zahar, 2005.

BIGLIARDI, R. V.; CRUZ, R. G. The role of environmental education in the face of the current civilization crisis. **Ambiente & Educagao**, Rio

Grande, v. 12, p. 127-141, 2007.

BOFF, L. **Sustainability:** what it is, what it isn't. Rio de Janeiro: Vozes, 2012.

BRAZIL. Ministry of Health. National Health Council. **Resolution no. 196, of October 10, 1996,** which deals with the approval of guidelines and regulatory standards for research involving human beings.

BUARQUE, S. C. **Metodologia de planejamento do desenvolvimento local e municipal sustentavel**. Brasilia: Inter-American Institute for Cooperation on Agriculture, 1999.

CACCIA BAVA, S. **The continuing search for identity**. 2007. Available at: <http://www.polis.org.br/artigo_interno.asp?codigo=157>. Accessed
on: October 29, 2013.

CARDOSO, D. M.; FREITAS, N. M.; TEIXEIRA, T. R. A.; SANTOS, S. P.; COSTA, K. S.; LISBOA, I. B. C.; CHAVES, G. P.; SANTOS, D. C. Education for sustainability: a multidisciplinary approach. In: Encontro da Sociedade Brasileira de Sociologia da Regiao Norte, 2, September 13-15, 2010, Belem. **Electronic Proceedings ....** Disponivel em: <http://www.sbsnorte2010.ufpa.br/site/anais/ARQUIVOS/GT14-110-481- 2010083118413.pdf. Accessed on: June 15, 2017.

CARNEIRO, D. M.; BAGOLIN, I. P. Approaches to the relationship between economic growth, inequality and poverty. **Revista Cadernos de Economia**, Chapeco, v. 16, n. 30-31, p. 5-20, 2012.

CARSON, R. Silent Spring. Sao Paulo: Gaia, 2010.

CAVALCANTI, C. Sustainability: mantra or moral choice? an ecological-economic approach. **Estudos Avangados**, Sao Paulo, v. 26, n.74, p. 35-50, 2012.

. Sustainability of the economy: Alternative paradigms of economic realization: In CAVALCANTI, C. (Org). **Development and nature**: studies for a sustainable society. 5. Ed - Sao Paulo: Cortez; Recife PE: Fundagao Joaquim Nabuco, 2009.

CECHIN, A. **Nature as the limit of the economy**. Sao Paulo: SENAC; EDUSP, 2010.

WORLD COMMISSION ON ENVIRONMENT AND DEVELOPMENT. **Our common future**. Rio de Janeiro: FGV, 1991.

DARWIN, E. **Temple of nature; or, origin of society**. **A poem**. 1804. Available at: < http://darwin-online.org.uk/converted/pdf/1804_Temple_A937.pdf. Accessed on: 13 Oct. 2016.

. **The loves of plants and animals**: romantic science and the pleasures of nature. 1794. Available at: <https://www.rc.umd.edu/>. Accessed on October 13, 2016. Accessed on October 13, 2016.

DIEGUES, A. C. **O mito moderno da natureza intocada.** Sao Paulo: Hucitec; Nucleo de Apoio a Pesquisa sobre Populapoes Humanas e Áreas Umidas Brasileiras, USP, 2000.

FONSECA NETO, S. N.; FREITAS, N. M. S.; FREITAS, N. M. S. A letter to the future: constructs on (in)sustainability. **Ciencia e Natura**, Santa Maria, v. 39 n.1,2017, p. 133-141, 2017.

FREITAS, J. **Sustentabilidade**: direito ao futuro. 2. ED. BELO HORIZONTE: FORUM, 2012.

FREITAS, N. M. S. **Desenvolvimento local e capital social**: uma analise interdisciplinar do processo de indupao do DLIS em Roraima. 2008. 296 f. Thesis (Doctorate). Postgraduate Program in Socio-environmental Development. Federal University of Para, 2008.

MARQUES, C. A. Convergences between the assumption of sustainability and the CTS approach in science education: an analytical experience in the context of chemistry teaching. In: Encontro Nacional de Ensino de Quimica, 18. Florianopolis, Santa Catarina, 2016. **Electronic proceedings...** Available at: <http://www.eneq2016.ufsc.br/an is/resumos/R1819-1.pdf>. Accessed on: Jan. 14, 2017.

GEORGESCU-ROEGEN, N. **Degrowth**: entropy, ecology, economy. GRINEVALD, J.; RENS, I. (Org.). ISAAC, M. J. P. (Trad.). Sao Paulo: SENAC, 2012.

GODOY, A. S.. Qualitative research: fundamental types. **Revista de**

Administrapao de Empresas, Sao Paulo, v. 35, n. 3, p. 20-29, 1995.

JACOBI, P. Educapao ambiental, cidadania e sustentabilidade. **Cadernos de Pesquisa**, Sao Paulo, n.118, p. 189-206, 2003.

KLIKSBERG, B. **Inequality in Latin America:** the debate postponed. Sao Paulo: Cortez; Brasilia, DF: UNESCO, 2002.

LAURIE, R. et al. Contributions of Education for Sustainable Development (ESD) to quality education: a synthesis of research. **Journal of Education for Sustainable Development**, USA, v. 10, n. 2, p. 226-242, 2016.

LEAR, L. Introduction. In: CARSON, Rachel. **Silent Spring**. Sao Paulo: Gaia, 2010. p. 11-19.

LEFF. E. **Betting for life**: sociological imagination and social imaginaries in the environmental territories of the South. Petropolis, RJ: Vozes, 2016.

. Complexity, environmental rationality and the dialog of knowledge. **Educapao e Realidade**, Porto Alegre, v. 34, n. 3, p. 17-24, 2009.

. **Epistemologia Ambiental**. 2 ed. Sao Paulo: Cortez, 2002.

LIMA, G. F. C. Critical environmental education: from socio-environmentalism to sustainable societies. **Educapao e Pesquisa**, Sao Paulo, v. 35, n. 1, p. 145-163, 2009.

LAYRARGUES, P. P. The cynicism of recycling: the ideological meaning of aluminum can recycling and its implications for environmental education. In: Loureiro, C. F. B; .; CASTRO, R. S. **Educapao ambiental**: repensando o espapo da cidadania (Orgs.). Sao Paulo Cortez, 2002. p. 179-220.

; LIMA, G. F. C. The political-pedagogical macro-trends of Brazilian environmental education. **Ambiente & Sociedade**, Sao Paulo, v. XVII, n. 1, p. 23-40, 2014.

LOBERA, J. Insostenibilidad: aproximacion al conflicto socioecologico. **Revista CTS**, Spain, n. 11, v. 4, p. 53-80, jul. 2008.

MARQUES, C. A. Understandings of chemical researchers about environmental sustainability. **Revista Brasileira de Ensino de Quimica**, Campinas, SP: Editora Atomo, v. 9 n. 2, p. 79-92, 2014.

MARQUES, L. **Capitalism and environmental collapse**. Campinas, SP: Editora da Unicamp, 2016.

MEADOWS, D. H. et al. **Limits to growth**: a report for the Club of Rome project on the dilemma of humanity. Sao Paulo: Editora Perspectiva, 1973.

MELO, M. M. **Capitalism versus sustainability**: the challenge of a new environmental ethic. Florianopolis: Editora da UFSC, 2006.

MELLO, L. F. Population, consumption and climate change. In: HOGAN, D. J.; MARANDOLA JUNIOR, E. **Population and climate change:** human dimensions of global environmental change (pp. 109-135). Sao Paulo: Nucleo de Estudos de Populapao (NEPO), 2009.

MINAYO, M. C. de S. O desafio da pesquisa social. In: (Org.). **Pesquisa social**: teoria, metodo e criatividade. 27. ed. Petropolis, RJ: Vozes, 2008. p. 9-29.

MORAES, R. Discursive dives: qualitative textual analysis understood as an integrated process of learning and inferring discourses. In: GALIAZZI, M. do C.; FREITAS, J. V. de (Org.). **Metodologias emergentes** de **pesquisa em educapao ambiental.** 2. ed. Ijui: Unijui, 2007. p. 85-114.

NABAES, T. O.; PEREIRA, V. A. Environmental Ontology: the repositioning of being on the horizon of environmental rationality. **Educar em Revista**, Curitiba, n. 61, p. 189-204, 2016.

NASCIMENTO, E. P. The trajectory of sustainability: from environmental to social, from social to economic. **Estudos Avanpados**, v. 26, n. 74, p. 5164, 2012.

NAVARRO, F. M.; GARCIA, J. C. T. **Ciencia para el mundo contemporaneo:** guia de recursos didacticos. Canarias, Spain: Agencia Canaria de Investigation, Innovation y Sociedad de la Informacion, 2010.

OLIVEIRA, M. M. **How to do qualitative research**. 6. ed. Petropolis: Vozes, 2014

PEDUZZI, S. S.; VILCHES, A.; GIL-PEREZ, D. Science of Sustainability. **Caderno Brasileiro de Ensino de Fisica**, Santa Catarina, v. 31, n. 3, p. 489-492, 2014.

UNEP. United Nations Environment Program. **XII Simulation of International Organizations**. Study Guide. Natal: SOI, 2012.

PORTO-GONQALVES, C. W. **The globalization of nature and the nature of globalization**. Rio de Janeiro: Civilizagao Brasileira, 2012.

RATTNER, H. **A bridge to a sustainable society**. Sao Paulo: SENAC, 2012.

RICKLEFS, R. E. **The Economics of Nature**. 6ª ed. Rio de Janeiro: Guanabara Koogan, 2010.

SACHS, I. Preface. In: VEIGA, J. E. **Desenvolvimento sustentavel:** o desafio do seculo XXI. Rio de Janeiro: Garamond, 2005. p. 9-11.

. Rethinking economic growth and social progress: the scope of politics. In: ARBIX, G.; ZILBOVICIUS, Mauro; ABRAMOVAY, Ricardo (Org.). **Razoes e ficgoes do desenvolvimento**. Sao Paulo: UNESP; EDUSP, 2001. p. 155-163.

SANTOS, B. S. **A gramatica do tempo**: para uma nova cultura politica. Sao Paulo: Cortez, 2006.

SERRAO, M.; ALMEIDA, A.; CARESTIATO. **Sustainability**: a matter for all of us. Rio de Janeiro: SENAC, 2012.

SCHONS, S. M. The environmental issue and the condition of poverty. **Revista katalysis,** v. 15, p. 70-78, v. 15, 2012.

SEN, A. K. **Development as freedom**. Sao Paulo: Companhia das Letras, 2010.

SUNKEL, O. The sustainability of current development in Latin America. In: ARBIX, Glauco; ZILBOVICIUS, Mauro; ABROMOVAY, Ricardo. **Reasons and fiction of development**. Sao Paulo: UNESP;

EDUSP, 2001. p. 267-310.

SILVA, L. F. **Critical environmental education**. Jundiai: Paco Editorial, 2014.

TRISTAO, M. **A educagao ambiental na formacao de professores**: redes de saberes. Sao Paulo: Annablume; Vitoria; Facitec, 2004.

UNESCO. **United Nations Decade for Sustainable Development, 2005-2014: Final Document of the International Implementation Scheme**. Brasilia, DF: UNESCO, 2005.

VACARRO, Santo Gabriel. **Jorge Luis Borges:** a critic of language. 2013. 350 f. Thesis (Postgraduate Program in Literature). Federal University of Santa Catarina. Florianopolis. 2013.

VASCONCELOS, E. R; CONCEIQAO, L. C. S.; FREITAS, N. M. S. Ideas about sustainable development: science education and the CTS approach, possible articulations. **REMEA - Revista Eletronica do Mestrado em Educagao Ambiental**, v. 28, p. 205-224, 2012.

.; FREITAS, N. M. S. The sustainability paradigm and the CTS approach: mediations for science teaching. **AMAZONIA** - Revista de Educagao em Ciencias e Matematicas, v. 9, n. 1, p. 89-108, 2012.

; FERREIRA, D. T.; VALENTE, J. A. S.; FREITAS, N. M. S. Educating for social and environmental justice: what issues to think about in the context of teaching and training science teachers? **Revista Brasileira de Pesquisa em Educacao em Ciencias**, Sao Paulo, v. 14, n. 2, p. 245- 254, 2014.

VECCHIATTI, K. Tres fases rumo ao desenvolvimento sustentavel: do reducionismo a valorização da cultura. **Sao Paulo em Perspectiva**, Sao Paulo, v.18, n. 3, p. 90-95, 2004.

VEIGA, J. E. **Desenvolvimento sustentavel**: o desafio do seculo XXI. Rio de Janeiro: Garamond, 2005.

. **Sustainability:** the legitimization of a new value. Sao Paulo: SENAC, 2010.

VILCHES, A.; GIL-PEREZ, D. Science of sustainability: a new field of knowledge to which chemistry and chemical education are contributing. **Educacion Quimica**, v. 24, n. 2, p. 199-206, 2013.

;. Sustainability Science: ^A new discipline or a new approach for all disciplines? **Revista Iberoamericana de Educación**, Spain, v. 69, n. 1, p. 39-60, 2015.

El Antropoceno como oportunidad para reorientar el comportamiento humano y construir un futuro sostenible. **Revista Electronica de Ensenanza de las Ciencias**, Spain, v. 10, n. 3, p. 394-419, 2011.

;. Planetary emergency: the need for a global plan. **Educatio Siglo XXI**, Spain, n. 25, p. 19-50, 2007.

WATANABE-CARAMELLO, G.; STRIEDER, R. B.; GEHLEN, S. T. Desafios e possibilidades para a abordagem de temas ambientais em aulas de Fisica. **Revista Brasileira de Pesquisa em Educapao em Ciencias**, Sao Paulo, v. 12, n. 1, p. 205-222, 2012.

WORLDWATCH INSTITUTE. **State of the world 2012**: towards sustainable prosperity. Salvador, BA: Uma Ed, 2012.

ZACARIAS, R. "Consumer Society", ideology of consumption and the socio-environmental inequities of current patterns of production and consumption. In: LOUREIRO, C. F. B.; LAYRARGUES, P. P.; CASTRO, R. S. (Orgs.). **Rethinking environmental education**: a critical look. Sao Paulo: Cortez, 2009. p. 119 - 139.

ZAPATA, Tania. Local economic development: challenges, limits and possibilities. In: **Local development**: trajectories and challenges. Recife: Instituto de Assessoria para o Desenvolvimento Humano, 2005. p. 15-28.

ZIBETTI, D. W. **Agricultural insurance and sustainable development**. Curitiba: Jurua, 2006.

ZOTTIS, H. G. A. et al. Violence and sustainable development: the role of the University. **Saude e Sociedade**, Sao Paulo, v. 17, n. 3, p. 33-41, 2008.

Printed by Books on Demand GmbH, Norderstedt / Germany